Low-Cost Methods for Molecular Characterization of Mutant Plants

Bradley J. Till • Joanna Jankowicz-Cieslak •
Owen A. Huynh • Mayada M. Beshir •
Robert G. Laport • Bernhard J. Hofinger

Low-Cost Methods for Molecular Characterization of Mutant Plants

Tissue Desiccation, DNA Extraction and Mutation Discovery: Protocols

Bradley J. Till
Plant Breeding and Genetics Laboratory
Joint FAO/IAEA Division of Nuclear
  Techniques in Food and Agriculture
Vienna
Austria

Owen A. Huynh
Plant Breeding and Genetics Laboratory
Joint FAO/IAEA Division of Nuclear
  Techniques in Food and Agriculture
Vienna
Austria

Robert G. Laport
School of Biological Sciences
University of Nebraska-Lincoln
Lincoln
Nebraska
USA

Joanna Jankowicz-Cieslak
Plant Breeding and Genetics Laboratory
Joint FAO/IAEA Division of Nuclear
  Techniques in Food and Agriculture
Vienna
Austria

Mayada M. Beshir
Agricultural Research Corporation
Khartoum North
Sudan

Bernhard J. Hofinger
Plant Breeding and Genetics Laboratory
Joint FAO/IAEA Division of Nuclear
  Techniques in Food and Agriculture
Vienna
Austria

ISBN 978-3-319-38667-6   ISBN 978-3-319-16259-1   (eBook)
DOI 10.1007/978-3-319-16259-1

Springer Cham Heidelberg New York Dordrecht London

Printed on acid-free paper

Springer International Publishing AG Switzerland is part of Springer Science+Business Media
(www.springer.com)

Foreword

The Joint FAO/IAEA Programme of Nuclear Techniques in Food and Agriculture has, for over 50 years, supported Member States in the use of nuclear techniques for crop improvement. This includes the use of induced mutations to generate novel diversity for breeding crops with higher yield, better nutritive value, and stronger resilience to biotic and abiotic stresses. This approach, first applied in the late 1920s, has been very successful across the world. More than 3,200 officially registered mutant crop varieties can be found in the IAEA's Mutant Variety Database. Covering over 150 species, examples include salt-tolerant rice, barley that can be grown at over 3,000 m, and wheat that is resistant to the emerging global disease known as Ug99. While successful, there are factors that threaten global food production and security. These include increasing world population and climate change and variation. Thus, continued and increasing efforts are required of plant breeding and genetics to meet the demand. Established and emerging biotechnologies that leverage available genome sequences can be used to facilitate and speed-up the plant breeding process. While successfully applied in developed countries, technology transfer to developing countries can be challenging. Issues include equipment and material costs and ease of experimental execution. The methods described in this book address this by providing low-cost and simple to execute molecular assays for germplasm characterization that can be applied in any laboratory equipped for basic molecular biology.

The views expressed in this text do not necessarily reflect those of the IAEA or FAO, or governments of their Member States. The mention of names of specific companies or products does not imply an intention to infringe on proprietary rights, nor should it be construed as an endorsement or recommendation on the part of IAEA or FAO.

Vienna, Austria   Bradley J. Till

Acknowledgements

We thank the participants of IAEA TC-funded training courses for their useful feedback when using these protocols during trainings held in the IAEA laboratories in Seibersdorf, Austria, from 2009 to 2014. We also thank Dr. Huijun Guo of the Chinese Academy of Agricultural Sciences for her assistance in evaluating an early draft of the DNA extraction protocol. We thank Dr. Thomas H. Tai of the United States Department of Agriculture and Dr. Jochen Kumlehn of the Leibniz Institute of Plant Genetics and Crop Plant Research, Germany, for serving as external reviewers and helping to improve this protocol book. Funding for this work was provided by the Food and Agriculture Organization of the United Nations and the International Atomic Energy Agency through their Joint FAO/IAEA Programme of Nuclear Techniques in Food and Agriculture. This work is part of IAEA Coordinated Research Project D24012.

Contents

Chapter 1
Introduction

Abstract A range of molecular methods can be employed for the characterization of natural and induced nucleotide variation in plants. These facilitate a better understanding of gene function and allow a reduction in the time needed to breed new mutant varieties. Molecular biology, however, can be difficult to master, and while efficient, many protocols rely on expensive pre-made kits. The FAO/IAEA Plant Breeding and Genetics Laboratory (PBGL) has developed a series of low-cost and easy to use approaches for the molecular characterization of mutant plant materials. The protocols are designed specifically to avoid complicated procedures, expensive equipment, and the use of hazardous chemicals. Furthermore, these protocols have been validated by research fellows from many developing countries.

1.1 Background

The extraction of high quality and quantity genomic DNA from tissues is at the heart of many molecular assays. Indeed, with the routine use of molecular markers and more recently the application of next generation sequencing approaches to characterize plant variation, the recovery of DNA can be considered a fundamental objective of the plant scientist, and is often a bottleneck in genotyping. The basic steps of DNA extraction are: (1) proper collection and storage of plant tissues, (2) lysis of plant cells, (3) solubilization of lipids and proteins with detergents, (4) separation of DNA from other molecules, (5) purification of the separated DNA, and (6) suspension in an appropriate buffer. Isolation of DNA dates to the late 1800s with the work of Friedrich Miescher and colleagues who first discovered the presence of DNA in cells long before it was established that DNA was the genetic material (Dahm 2005).

B.J. Till et al., *Low-Cost Methods for Molecular Characterization of Mutant Plants*,
DOI 10.1007/978-3-319-16259-1_1

1.2 Methods Used to Isolate Genomic DNA from Plant Tissues

The advent of recombinant DNA technologies and DNA sequencing technologies in the 1970s marked the beginning of a rapid expansion of molecular biology analyses in plants that continues to this day. In parallel, DNA isolation procedures tailored to the unique aspects of plant cells have evolved. A range of DNA extraction methods have been described; however, some are more commonly used by plant biologists. One of the most enduring methods for plant DNA extraction employs a lysis buffer the main component of which is cetyltrimethy-lammonium bromide (CTAB), which solubilizes membranes and complexes with the DNA. The so-called CTAB method, first described in 1980, employs an organic phase separation using a chloroform-isoamyl alcohol extraction, and alcohol pre-cipitation to isolate DNA from proteins and other materials (Murray and Thompson 1980). The method remains popular in part due to the fact that all components can be self-prepared, and thus the per-sample cost remains low. Wide and prolonged usage of the method also validates the approach for many different molecular assays. However, manual phase separation means that human error can introduce unwanted cross-contamination of organic compounds that may result in an inhibi-tion of downstream enzymatic assays. Further, chloroform is a toxic organic compound and proper ventilation and waste disposal measures are needed.

An alternative to the CTAB method is the use of high concentrations of potas-sium acetate and the detergent sodium dodecyl sulfate (SDS) (Dellaporta et al. 1983). Proteins and polysaccharides are precipitated and removed from the soluble DNA. This approach is advantageous to the CTAB method in that organic phase separation is avoided. An additional filtration step may be required to remove cell wall debris and other insoluble materials from soluble DNA, limiting the through-put of the method.

In recent decades, commercial kits for the rapid extraction of DNA from plant tissues have been routinely used by many laboratories. Commercial kits have proven to be very reliable in producing high yields of highly purified DNA and so have become the standard when performing sensitive molecular assays. Many such kits utilize the binding of DNA to silica in the presence of chaotropic salts. In the presence of high concentrations of chaotropic salt, the interaction of water with the DNA backbone is disrupted and charged phosphate on the DNA can form a cationic bridge with silica, while other components remain in solution. Silica is either used in a solid phase as with spin columns, or in a slurry form for batch chromatography. Washing the DNA-bound silica in the presence of a high percent-age of alcohol removes excess salt. The subsequent addition of an aqueous solvent (water or buffer) drives the hydration of the DNA and its subsequent release from silica. The now soluble DNA can be separated from silica through a quick centri-fugation step. The method is rapid, taking less than 1 h, and is scalable such that a 96-well plate format is commonly employed to increase sample throughput. While highly advantageous over other methods, such kits remain expensive when compared to home-made ones such as the CTAB and Dellaporta protocols. Therefore, the methods described here were developed to provide the ease and quality of silica-based DNA extraction at a fraction of the cost while using basic laboratory equipment.

1.3 Methods for the Discovery and Characterization of Induced and Natural Nucleotide Variation in Plant Genomes

Nucleotide variation is the major source of the phenotypic diversity that is exploited by plant breeders. Variation can be either natural or induced. In the late 1990s, a reverse-genetic strategy was developed whereby induced mutations were used in combination with novel methods for the discovery of nucleotide variation (McCallum et al. 2000). Known as Targeting Induced Local Lesions IN Genomes (TILLING), this approach allows for the recovery of multiple new alleles in any gene in the genome, provided the correct balance of population size and mutation density can be achieved (Colbert et al. 2001; Till et al. 2003). Efficient techniques for the discovery of Single Nucleotide Polymorphisms (SNP) and small insertion/deletions (indels) were developed utilizing single-strand-specific nucleases that can be easily prepared through extractions of plants such as celery, or mung beans (Till et al. 2004). TILLING has been applied to over 20 plant and animal species, and similar approaches have been used to characterize naturally occurring nucleotide variation, known as Ecotilling (Comai et al. 2004; Jankowicz-Cieslak et al. 2011). While TILLING and Ecotilling have been primarily used in seed crops, the methods work well in vegetatively (clonally) propagated and polyploid species such as banana and cassava (Jankowicz-Cieslak et al. 2012; Till et al. 2010). The PBGL has developed low-cost methods for the extraction of enzymes from a variety of plant materials, including easily obtainable weedy plants. The laboratory has also adapted low-cost agarose gel-based TILLING and Ecotilling assays.

References

Colbert T, Till BJ, Tompa R, Reynolds S, Steine MN et al (2001) High-throughput screening for induced point mutations. Plant Physiol 126:480–484
Comai L, Young K, Till BJ, Reynolds SH, Greene EA et al (2004) Efficient discovery of DNA polymorphisms in natural populations by Ecotilling. Plant J 37:778–786
Dahm R (2005) Friedrich Miescher and the discovery of DNA. Dev Biol 278:274–288

Dellaporta SL, Wood J, Hicks JB (1983) A plant DNA minipreparation: version II. Plant Mol Biol
    Rep 1:19–21
Jankowicz-Cieslak J, Huynh OA, Bado S, Matijevic M, Till BJ (2011) Reverse-genetics by
    TILLING expands through the plant kingdom. Emir J Food Agric 23:290–300
Jankowicz-Cieslak J, Huynh OA, Brozynska M, Nakitandwe J, Till BJ (2012) Induction, rapid
    fixation and retention of mutations in vegetatively propagated banana. Plant Biotechnol J
    10:1056–1066
Mccallum CM, Comai L, Greene EA, Henikoff S (2000) Targeted screening for induced muta-
    tions. Nat Biotechnol 18:455–457
Murray MG, Thompson WF (1980) Rapid isolation of high molecular weight plant DNA. Nucleic
    Acids Res 8:4321–4325
Till BJ, Reynolds SH, Greene EA, Codomo CA, Enns LC et al (2003) Large-scale discovery of
    induced point mutations with high-throughput TILLING. Genome Res 13:524–530
Till BJ, Burtner C, Comai L, Henikoff S (2004) Mismatch cleavage by single-strand specific
    nucleases. Nucleic Acids Res 32:2632–2641
Till BJ, Jankowicz-Cieslak J, Sagi L, Huynh OA, Utsushi H et al (2010) Discovery of nucleotide
    polymorphisms in the Musa gene pool by Ecotilling. Theor Appl Genet 121:1381–1389

Chapter 2
Health and Safety Considerations

Abstract All laboratories should have standardized health and safety rules and practices. These can vary from region to region due to differences in legislation. Before beginning new experiments, please consult your local safety guidelines. Failure to follow these rules could result in accidents, fines, or a closure of the laboratory. Consider the following guidelines in this chapter applicable to all laboratories.

More information on general laboratory practices is available (Barker 2005).

2.1 Guidelines

1. Always wear a laboratory coat in the laboratory. Remove the coat when exiting the lab to avoid contaminating people with the things you are protecting yourself from.
2. Wear eye protection (special safety goggles) when working with chemicals or anything that you don't want entering your eye.
3. Wear gloves to protect your hands from dangerous materials, and to protect your samples from contamination. Standard laboratory gloves made of latex or nitrile are suitable for the methods described. Powder-free gloves are advised when using equipment with precision optics. Do not touch common items like the telephone, door handles, or light switches with gloves as the next person touching those items may not be protected from hand contamination. The same rule applies to mobile phones. Remove gloves before leaving the laboratory.
4. Wear proper foot protection, and avoid open toe footwear and high heels.
5. Wear clothing that covers your legs. Avoid loose fitting clothing that can be caught in machinery or be passed over an open flame.
6. Familiarize yourself with emergency procedures. Know where the nearest eye-wash station and shower are located. Know where the nearest first aid kit is located, and locate the list of emergency telephone numbers.
7. Consult the Materials Safety Data Sheet (MSDS) for the chemicals you will be using. These sheets should come with the chemicals. They provide information

© International Atomic Energy Agency 2015
B.J. Till et al., *Low-Cost Methods for Molecular Characterization of Mutant Plants*,
DOI 10.1007/978-3-319-16259-1_2

on health risks, first aid measures, fire and explosion data, how to deal with accidental release (spills), handling and storage, and guidelines for personal protection. If you don't have the MSDS, you can find them by doing a web search of the item with MSDS in the title. Note that it is a best practice to review the MSDS supplied by the manufacturer of the chemical you have in your own laboratory. Similar chemical names or other formulations may result in misleading web search results.

8. Locate the emergency spill kit to handle accidental spillage of hazardous materials. If your laboratory is not equipped, consider preparing one (see Sect. 2.2).

9. Don't rush. If you are unfamiliar with a piece of equipment, or concerned about the safety of a procedure, stop! Make sure you know what you are doing and the risks associated with the procedures before you begin. Many laboratories use a written standard operating procedure (SOP) that is followed during the initial performance of a protocol or procedure and made available for future reference. Check with the procedures of your laboratory and consider employing an SOP approach.

2.2   Preparation of a Home-Made Chemical Spill Kit

All laboratories should contain a kit for chemical spills. While spill kits are commercially available, self-prepared ones can be made at a fraction of the cost. Key materials and their use are found in Table 2.1. The kit should be designed to handle a spill from the largest volume of chemical you have in the laboratory. For

Table 2.1 Components of a chemical spill kit and their uses[a]

Component	Use
Five gallon plastic or rubber bucket with lid clearly labelled "Chemical Spill Kit" with emergency telephone numbers printed clearly on the lid and the side of the bucket	This bucket contains all the materials of the spill kit, and should be located near the laboratory doorway to allow someone to access it after they have left the spill area
Goggles	For eye protection while cleaning spill
Chemical-resistant gloves	For hand protection when dealing with spills
Absorbent materials (cat litter, vermiculite, activated charcoal, or sawdust)	This material is placed on the liquid spills to contain the liquid for easy removal
Small broom and plastic dustpan	For removal of dry spills, and absorbed materials. It is important that the dustpan or scoop be plastic as metal materials can spark and cause fire/explosions
Sturdy plastic bags	To contain materials
Baking soda (sodium bicarbonate), in a plastic bag marked "for liquid acid spills"	For neutralization of small acid spills
Acetic acid powder in a plastic bag marked "for liquid base spills"	For neutralization of small base spills

[a]Note that material collected after a spill should not go into the normal waste but be disposed of in the appropriate manner according to the local guidelines for hazardous waste

detailed guidelines please refer to the "Guide for Chemical Spill Response Planning in Laboratories" prepared by the American Chemical Society (http://www.acs.org/content/acs/en/about/governance/committees/chemicalsafety/publications/guide-for-chemical-spill-response.html).

Reference

Barker K (2005) At the bench: a laboratory navigator. Cold Spring Harbor Press, New York, NY

Chapter 3
Sample Collection and Storage

Abstract Of importance to the successful extraction of genomic DNA from plant tissues is the collection of the suitable material and proper storage of the tissues before DNA isolation. If the samples are not properly treated, DNA can be degraded prior to isolation. The rate of sample degradation can vary dramatically from species to species depending on the method of sample collection. Mechanisms of genomic DNA degradation include exposure to endogenous nucleases due to organellar and cellular lysis. To prevent this from occurring, leaf or root tissues are commonly flash frozen in liquid nitrogen and then stored at −80 °C. At these temperatures, nucleases remain inactive and DNA is stable. Thawing of tissue in some species can lead to rapid degradation. Therefore, during the extraction procedure, it may be necessary to grind the tissue to a fine powder in the presence of liquid nitrogen and expose frozen tissue immediately to a lysis buffer containing EDTA, which inhibits nuclease activity. This chapter provides an alternative method for sample collection and storage. Silica gel is used to desiccate tissues at room temperature. This avoids the use of liquid nitrogen and storage at −80 °C.

3.1 Background

While collection of tissues in liquid nitrogen and −80 °C storage may be highly suitable for most plant species, it can be impractical in some developing countries owing to the expense and difficulty in procuring liquid nitrogen. The provision of continuous power supplies for ultralow (−80 °C) freezers may also be difficult and costly. Lyophilization, or freeze drying, is an alternative approach that results in tissue samples that can be stored at room temperature for many months prior to the isolation of DNA. This has been used to produce high quality genomic DNA suitable for high throughput TILLING assays (Till et al. 2004). Lyophilization circumvents the need for continual −80 °C storage, but commercial lyophilizers are also expensive. An alternative method is described in this chapter. Tissue is collected and stored in silica gel (Chase and Hills 1991; Liston et al. 1990). This removes water from tissues, and in many cases the dried material is stable at room temperature for weeks to months before the isolation of DNA. The exact length of

© International Atomic Energy Agency 2015

B.J. Till et al., *Low-Cost Methods for Molecular Characterization of Mutant Plants*,
DOI 10.1007/978-3-319-16259-1_3

Table 3.1 Materials for collection, storage, and desiccation of plant tissues

Material description	Examples of suppliers and catalogue numbers
Scissors	Any supplier
Porous paper envelopes	Any supplier
Silica gel with moisture indicator	Sigma 13767
Container for storing tissue with silica gel	Any supplier
Parafilm® for sealing container	Sigma P7793

time that dried tissue can be stored and still yield suitable quantities and quality of genomic DNA should be determined empirically. Other factors such as stress-induced accumulation of phenolic compounds may also limit the utility and shelf-life of the material. This is likely to vary between species and genotypes (Savolainen et al. 1995).

3.2   Materials

Materials needed for the desiccation of plant tissues at room temperature are listed in Table 3.1.

3.3   Methods

1. Label envelopes for tissue storage. Tissue desiccation works best when it is stored in porous materials. Paper envelopes, tea bags, or kimwipes work well.
2. The material should be cut to roughly the same length as the collection envelope to facilitate desiccation (Fig. 3.1, left panel).
3. Immediately upon collection, place the envelopes containing the leaf material into a container containing silica gel. Seal the container with Parafilm to limit the effects of atmospheric humidity. The ratio of silica gel to tissue should be no less than 10:1 by weight (Weising et al. 2005). Orange silica gel has a moisture indicator. When fully dehydrated and ready for use, it is orange; when fully hydrated, the silica gel turns white (Fig. 3.1, right panel). The silica gel can be dehydrated by heating at a high temperature (over 80 °C) until the color returns to orange and may be re-used many times.
4. Incubate the material with silica gel for at least 48 h at room temperature (RT). The tissue is suitable for DNA extraction when brittle. Incubate for additional time if necessary. The tissue can be stored for long periods (>1 month) in silica gel at RT. It is suggested that you perform the tests in your own laboratory to determine the maximal amount of time that tissue can be stored under these conditions.

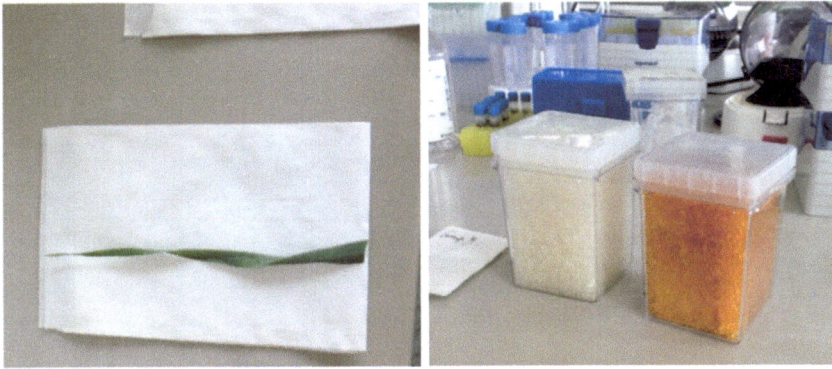

Fig 3.1 Leaf tissue is collected and placed in an envelope (*left panel*). The length of the tissue should equal the length of the envelope. Silica gel with color indicator turns *white* when fully hydrated (*right panel*). Only gel with an *orange color* should be used

References

Chase MW, Hills HH (1991) Silica gel: an ideal material for field preservation of leaf samples for DNA studies. Taxon 40:215–220

Liston A, Rieseberg LH, Adams RP, Do N, Zhu G (1990) A method for collecting dried plant specimens for DNA and isozyme analyses, and the results of a field experiment in Xinjiang, China. Ann Missouri Bot Gard 77:859–863

Savolainen V, Cuenoud P, Spichiger R, Martinez MD, Crevecoeur M et al (1995) The use of herbarium specimens in DNA phylogenetics: evaluation and improvement. Plant Syst Evol 197:87–98

Till BJ, Reynolds SH, Weil C, Springer N, Burtner C et al (2004) Discovery of induced point mutations in maize genes by TILLING. BMC Plant Biol 4:12

Weising K, Nybom H, Wolff K, Kahl G (2005) DNA fingerprinting in plants: principles, methods and applications. CRC Press, Boca Raton, FL

Chapter 4
Low-Cost DNA Extraction

Abstract The methods described in this chapter were developed to avoid toxic organic phase separation utilized in many low-cost DNA extraction protocols such as the CTAB method. The steps involve: (1) lysis of the plant material, (2) binding of DNA to silica powder under chaotropic conditions, (3) washing the bound DNA, and (4) elution of DNA from the silica powder. This method has been tested in several plant species and the applicability of such DNA preparations for molecular marker studies in barley is shown in Chap. 8.

4.1 Materials

Chemicals, enzymes, and equipment are listed in Table 4.1, and working stocks in Table 4.2.

4.2 Methods

4.2.1 Preparation of Silica Powder DNA Binding Solution

1. Transfer silica powder (Celite 545 silica) into a 50-ml conical tube (fill to the 2.5 ml line = approximately 800 mg).
2. Add 30 ml dH$_2$O.
3. Shake vigorously (vortex and invert 15 times or until a hydrated slurry forms).
4. Let the slurry settle for approximately 15 min.
5. Remove (pipette off) the liquid.
6. Repeat steps 2–5 an additional two times for a total of three washes. After the last washing step suspend the hydrated silica in a volume of water equal to the volume of silica (typically up to the 5-ml mark on the conical tube). This is the liquid silica stock (LSS) and can be stored at RT for up to 1 month.
7. Prior to use, suspend LSS by vortexing for 30 s or until a homogenized slurry is formed. Transfer 50 µl into 2-ml tubes. Prepare one tube per tissue sample.

© International Atomic Energy Agency 2015

B.J. Till et al., *Low-Cost Methods for Molecular Characterization of Mutant Plants*,
DOI 10.1007/978-3-319-16259-1_4

Table 4.1 Chemicals, enzymes and equipment for low-cost DNA extraction

Material description	Examples of suppliers and catalogue numbers
Celite 545 silica powder (Celite 545-AW reagent grade)	Sigma 20199-U
SDS (sodium dodecyl sulfate)	Sigma L-4390
50-ml conical tube with cap	Fisher Scientific 14-432-22
Sodium acetate anhydrous	Sigma S-2889
NaCl (sodium chloride)	Sigma S-3014
RNase A	Sigma R6513
Ethanol (absolute)	Fisher Scientific BP2818-4
H₂O (distilled or deionized and autoclaved)	
Potassium iodide	Sigma P2963
Guanidine thiocyanate (optional)	Sigma G9277
Microcentrifuge tubes (1.5 and 2.0 ml)	Any general laboratory supplier
Micropipettes (1,000, 200, and 20 µl)	Any general laboratory supplier
Microcentrifuge	Eppendorf Centrifuge 5415D
Vortex mixer	Vortex Genie 2, Fisher Scientific NC9864336
Metal beads (tungsten carbide beads, 3 mm)	Qiagen 69997
Sea sand (optional)	Sigma 274739
Agarose gel equipment	Horizontal electrophoresis from any general laboratory supplier

ATTENTION: try to keep the silica suspended when transferring to tubes to ensure that all tubes receive the same amount of LSS.

8. Add 1 ml H₂O per tube to perform a final wash step.
9. Mix by vortexing for 15 s or until silica is fully suspended.
10. Centrifuge at full speed ($16,000 \times g$) for 20 s.
11. Pipette off the liquid.
12. Add 700 µl DNA binding buffer (6 M potassium iodide or alternatively 6 M guanidine thiocyanate).
13. Suspend the silica in DNA binding buffer by vortexing for 15 s.
14. The Silica Binding Solution (SBS) is now ready for use.

4.2.2 Low-Cost Extraction of Genomic DNA

1. Prepare an ice bath.
2. Label 2-ml tubes containing three metal tungsten carbide beads with sample names.
3. Add the dried tissue to the appropriate tube.
4. Tape the tubes onto a vortex mixer (Fig. 4.1) and vortex on high setting for 30 s or until the material is ground to a fine powder. NOTE: If the tissue is not fully ground, grinding is facilitated by addition of 0.2 g of purified sea sand per tube. It is common for some tissues to not be completely ground to a powder. High

Table 4.2 Working stocks for DNA extraction

Stock solution	Recipe	Comments
5 M NaCl	MW = 58.44 g/mol 29.22 g/100 ml	Do not use if precipitate forms. Either heat to get fully back into solution or discard and make fresh
3 M Sodium Acetate (pH 5.2)	MW = 82.03 g/mol 24.61 g/100 ml	Adjust pH value with glacial acetic acid
95 % (v/v) Ethanol	95 ml ethanol abs 5 ml H_2O	Use fresh. Ethanol absorbs water and the % will drop over time
Tris-EDTA (TE) buffer (10×)	100 mM Tris-HCl, pH 8.0 10 mM EDTA	Tris and EDTA can be prepared from powders. This may be less costly. However, note that the pH of Tris changes with temperature
Lysis buffer	0.5 % SDS (w/v) in 10× TE 0.5 g SDS/100 ml	
DNA binding buffer	6 M Potassium Iodide (KI) Alternative: 6 M Guanidine thiocyanate	ATTENTION! It takes several hours until fully dissolved (leave it for approximately 4–5 h)
Wash buffer	1 ml of 5 M NaCl 99 ml of 95 % EtOH	ATTENTION! Prepare fresh because the salt precipitates during storage
DNA elution buffer	1x TE-buffer	Tris-EDTA buffer is advised for most applications. If the presence of EDTA is inhibitory to downstream applications, elution can be carried out using 10 mM Tris

Fig. 4.1 Sample grinding is accomplished by combining desiccated leaf material and metal beads into a 2-ml tube (*left panel*) and taping sample tubes to a standard vortex mixer (*middle panel*). Grinding is complete when a find powder is produced. The presence of unground tissue with the powder does not affect the quality of extracted DNA (*right panel*, and example data in Chap. 8)

quality DNA can still be produced from such samples. Sample degradation can occur after prolonged vortexing. It is therefore suggested to test different grinding times to find proper conditions to maximize both genomic DNA yield and quality.

5. Add 800 μl of Lysis Buffer and 4 μl RNAse A (10 μg/ml) to each tube. NOTE: See Sect. 4.3 for alternative buffers.

6. Vortex at a high speed for approximately 2 min until the powder is fully hydrated and mixed with buffer.
7. Incubate for 10 min at RT.
8. Add 200 µl 3 M sodium acetate (pH 5.2). Mix by the inversion of tubes and incubate on ice for 5 min.
9. Centrifuge at 16,000×g for 5 min at RT to pellet the leaf material.
10. Label the tubes with aliquots of silica binding solution (SBS, 700 µl) with the sample name.
11. Transfer the liquid into appropriately labelled SBS-containing tubes. DO NOT TRANSFER THE LEAF MATERIAL!
12. Completely suspend the silica powder by vortexing and inverting the tubes (approximately 20 s).
13. Incubate for 15 min at RT (on a shaker at 400 rpm, or invert tubes every 3 min by hand).
14. Centrifuge at 16,000×g for 3 min at RT to pellet the silica.
15. Remove the supernatant with a pipette and discard (the DNA is bound to the silica at this stage).
16. Add 500 µl of freshly prepared wash buffer to each tube.
17. Completely suspend the silica powder by vortexing or inverting the tubes (approximately 20 s).
18. Centrifuge at 16,000×g for 3 min at RT to pellet the silica. Remove the supernatant and keep the pellet.
19. Repeat steps 16–18.
20. Centrifuge the pellet for 30 s and remove any residual wash buffer with a pipette.
21. Open the lid on tubes containing silica pellet and place in a fume hood for 30 min to fully dry the pellets (NOTE: This can be done for a longer period on the bench top if a fume hood is not available).
22. Add 200 µl TE buffer to each tube to elute the DNA. The DNA is now in the liquid buffer. A buffered solution is preferred over water to prevent degradation.
23. Completely suspend the silica powder by vortexing and inversion of tubes (approximately 20 s).
24. Incubate at RT for 5 min.
25. Centrifuge at 16,000×g for 5 min at RT to pellet the silica.
26. Label new 1.5-ml tubes with sample numbers/codes.
27. Collect the liquid containing genomic DNA and place into new tubes.
28. Store DNA temporarily at 4 °C before checking the quality and quantity.
29. Evaluate the quality and quantity of the extracted DNA. While fluorometric and spectrophotometric methods have their advantages, it is suggested that samples are evaluated using agarose gel electrophoresis and a quantitative marker so that sample degradation and the presence of any RNA can be monitored. See Chap. 8 for example data.

Table 4.3 Alternative lysis buffers for DNA extraction

Lysis buffer (LB) name	Recipe (in 10× TE)	Preparation in final volume of 100 ml in 10× TE
LB1	0.5 % SDS (w/v)	0.5 g SDS
LB2	0.5 % SDS (w/v)	0.5 g SDS
	0.5 M NaCl	10 ml of 5 M NaCl
	3 % PVP (w/v)	3 g
LB3	0.5 % SDS (w/v)	0.5 g SDS
	0.5 M NaCl	10 ml of 5 M NaCl
	3 % PVP (w/v)	3 g
	1 % sodium sulfite	1 g
LB4	0.5 % SDS (w/v)	0.5 g SDS
	0.5 M NaCl	10 ml of 5 M NaCl
	3 % PVP (w/v)	3 g
	1 % sodium sulfite	1 g
	2 % N-lauryl-sarcosyl sodium salt	2 g

4.3 Alternative Buffers for DNA Extraction

The main areas for the optimization of DNA-extraction methods include increasing sample yield, reducing co-purification of unwanted components (e.g., polysaccharides, and polyphenols), and reducing sample degradation. To a large extent, providing the starting tissues are of good quality, all three areas can be influenced by the sample lysis procedure. Table 4.3 lists four lysis buffers to optimize the isolation of DNA from grapevine and sorghum. A more thorough compilation of buffer components and additives to enhance DNA isolation in the presence of secondary compounds can be found in Weising et al. (2005). Data from buffer optimizations are shown in Chap. 8.

Reference

Weising K, Nybom H, Wolff K, Kahl G (2005) DNA fingerprinting in plants: principles, methods and applications. CRC Press, Boca Raton, FL

Chapter 5
PCR Amplification for Low-Cost Mutation Discovery

Abstract PCR is used to amplify regions to be interrogated for the presence of mutations (SNP and small indel polymorphisms). While PCR is a common practice and many protocols exist, reaction conditions are provided here that are optimized for TILLING and Ecotilling assays utilizing native agarose gel electrophoresis.

5.1 Materials

Consumables and equipment for PCR amplification are listed in Table 5.1.

Table 5.1 Chemicals, enzymes, and equipment for PCR amplification

Material description	Examples of suppliers and specifications
Genomic DNA	Concentration 0.075 ng/µl for a 150-Mbp diploid genome. Scale accordingly and test different concentrations as necessary
TaKaRa HS Taq[a], 5 U/µl	ExTaq kit, TaKaRa, Japan
ExTaq PCR buffer	ExTaq kit, TaKaRa, Japan
dNTPs	ExTaq kit, TaKaRa, Japan
Forward and reverse primers	Tm 67–73 °C, designed to amplify a specific genomic region producing an amplicon between 800 and 1,600 bp. Primer design is aided with freely available software such as Primer3 (Rozen and Skaletsky 2000)
H_2O (distilled or deionized and autoclaved)	
DNA size ladder	Any standard ladder providing sizing standards between 100 bp and 2 kb, e.g., 1 kb Plus, Life Technologies 10787-018
0.2 ml tubes	Any general laboratory supplier
Thermocycler	Any standard thermocylcer, e.g., Biorad C1000 Thermal cycler
Microcentrifuge	Any standard microcentrifuge, e.g., Eppendorf Centrifuge 5415D
Agarose gel equipment	Horizontal electrophoresis from any general laboratory supplier

[a]While hot start Taq polymerases can offer improved results, lower cost polymerases can be used for PCR amplification

© International Atomic Energy Agency 2015
B.J. Till et al., *Low-Cost Methods for Molecular Characterization of Mutant Plants*,
DOI 10.1007/978-3-319-16259-1_5

5.2   Methods

1. Prepare a PCR master mix on ice by combining:

H₂O	82.5 μl
10× Ex Taq buffer	15 μl
2.5 mM dNTP mix	12 μl
10 μM L primer	1.5 μl
10 μM R primer	1.5 μl
TaKaRa HS taq (5 U/μl)	0.38 μl

2. Mix the PCR master mix by pipetting it up and down ten times followed by pulse centrifugation.
3. Combine 7.5 μl DNA at the appropriate concentration with 22.5 μl of PCR master mix. Mix by pipetting it up and down.
4. Incubate in a thermal cycler using the following parameters:
     95 °C for 2 min; loop 1 for 8 cycles (94 °C for 20 s, 73 °C for 30 s, reduce temperature 1 °C per cycle, ramp to 72 °C at 0.5 °C/s, 72 °C for 1 min); loop 2 for 45 cycles (94 °C for 20 s, 65 °C for 30 s, ramp to 72 °C at 0.5 °C/s, 72 °C for 1 min); 72 °C for 5 min; 99 °C for 10 min; loop 3 for 70 cycles (70 °C for 20 s, reduce temperature 0.3 °C per cycle); hold at 8 °C.
5. OPTIONAL: Check the yield of the PCR product by agarose gel electrophoresis. See Chap. 8 for example data. For the efficient discovery of nucleotide polymorphisms, PCR product yield should be approximately 10 ng/μl or higher in concentration. PCR product should be a single band. Co-amplification of multiple sequences can result in high error rates (Cooper et al. 2008).

References

Cooper JL, Till BJ, Laport RG, Darlow MC, Kleaffner JM et al (2008) TILLING to detect induced mutations in soybean. BMC Plant Biol 8:9

Rozen S, Skaletsky H (2000) Primer3 on the WWW for general users and for biologist pro grammers. In: Krawetz S, Misener S (eds) Methods in molecular biology. Humana Press, Totowa, NJ, pp 365–386

Chapter 6
Enzymatic Mismatch Cleavage and Agarose Gel Evaluation of Samples

Abstract Denaturation and annealing of PCR products allows DNA strands with small sequence differences to hybridize together. The result is heteroduplexed molecules that are single stranded in polymorphic sequence locations, but double stranded elsewhere. These molecules are the substrates for cleavage by single-strand-specific nucleases such as CEL I, crude Celery Juice Extract (CJE) containing CEL I, and other plant extracts containing single-strand-specific nucleases [Till et al. (Nucleic Acids Res, 32:2632–2641, 2004)]. Enzymatic cleavage initiates on a single strand and can result in double strand breaks. The products of cleavage can therefore be observed using native gel electrophoresis.

6.1 Materials

Consumables and equipment for enzymatic mismatch cleavage are listed in Table 6.1.

6.2 Methods

1. Prepare the following enzyme master mix on ice (calculated for five samples):
   81.5 µl water
   15 µl 10× CEL I buffer
   3.5 µl CJE nuclease
2. Label four new PCR tubes with the sample name.
3. Combine 20 µl of PCR product with 20 µl of enzyme master mix. Pipette the mixture up and down to mix or vortex briefly followed by pulse centrifugation.
4. Incubate at 45 °C for 15 min in a thermal cycler.
5. Place the reactions on ice, stop the reaction by adding 10 µl of 0.25 M EDTA per sample, and mix well by vortexing and centrifuge briefly (NOTE: Samples can be stored frozen for months before analysis).

© International Atomic Energy Agency 2015
B.J. Till et al., *Low-Cost Methods for Molecular Characterization of Mutant Plants*,
DOI 10.1007/978-3-319-16259-1_6

Table 6.1 Chemicals, enzymes, and equipment for enzymatic mismatch cleavage

Material description	Examples of suppliers and specifications
10× CELI buffer	5 ml 1 M MgSO$_4$, 100 μl 10 % Triton X-100, 5 ml 1 M Hepes (pH 7.4), 5 μl 20 mg/ml bovine serum albumin, 2.5 ml 2 M KCl, 37.5 ml water
Crude Celery Juice Extract (CJE)	See Till et al. (2004) for the preparation of enzyme and defining unit activity. Chap. 7 provides a protocol for the preparation of single-strand-specific nucleases from weedy plants
1 kb DNA ladder	Any general laboratory supplier
0.25 M EDTA	Prepared from ethylenediaminetetraacetic acid (EDTA) stock from any general laboratory supplier
H$_2$O	Distilled or deionized and autoclaved
1.5 ml, 2.0 ml tubes	Any general laboratory supplier
Thermocycler	e.g., Biorad C1000 Thermal cycler
Microcentrifuge	Eppendorf Centrifuge 5415D
Agarose gel equipment	Horizontal electrophoresis from any general laboratory supplier

6. Analyze the samples by electrophoresis using a 1.5 % agarose gel. See Chap. 8 for example data.

Open Access This chapter is distributed under the terms of the Creative Commons Attribution Noncommercial License, which permits any noncommercial use, distribution, and reproduction in any medium, provided the original author(s) and source are credited.

Reference

Till BJ, Burtner C, Comai L, Henikoff S (2004) Mismatch cleavage by single-strand specific nucleases. Nucleic Acids Res 32:2632–2641

Chapter 7
Alternative Enzymology for Mismatch Cleavage for TILLING and Ecotilling: Extraction of Enzymes from Common Weedy Plants

Abstract A crude celery extract containing the single-strand-specific nuclease CEL I, has been widely used in TILLING and Ecotilling projects around the world. Yet, celery is hard to come by in some countries. Sequences homologous to CEL I can be found in different plant species. Previous work showed that similar mismatch cleavage activities could be found in crude extracts of mung bean (Till BJ, Burtner C, Comai L, Henikoff S. Nucleic Acids Res 32:2632–2641, 2004). It is likely that the same activity can be recovered in many different plant species. Therefore, a protocol for the extraction of active enzyme was developed that uses plants common across the world, namely weeds. Monocotyledenous and dicotyledenous weedy plants from the grassland, field and waste grounds around crop fields are suitable for this protocol. Due to lower recovery of enzymatic activity compared to celery-based extractions, a centrifuge-based filter method is applied to concentrate the enzyme extract.

7.1 Materials

Extraction of single-strand-specific nuclease from weedy material is performed using standard laboratory equipment and consumables (Table 7.1). Concentration of enzyme extracts is accomplished using specialized centrifugation filters (Table 7.2), and testing of enzyme activity relies on standard materials for PCR (Table 7.3).

7.2 Methods

7.2.1 Enzyme Extraction

1. Collect approximately 200 g of mixed monocot and dicot weedy plants. Wash material 3× in water and then grind using a hand-held mixer and by adding about 300 ml of water to facilitate tissue disruption.

© International Atomic Energy Agency 2015

B.J. Till et al., *Low-Cost Methods for Molecular Characterization of Mutant Plants*,
DOI 10.1007/978-3-319-16259-1_7

Table 7.1 Chemicals, enzymes, and equipment for extraction of enzymes from common weedy plants

Material description	Comment
Hand-held mixer (or juicer)	From any supplier
STOCK: 100 mM phenylmethylsulfonyl fluoride (PMSF; stock in isopropanol)	To prepare an aqueous solution of 100 µM PMSF (for buffers A and B), add 1 ml 0.1 M PMSF per liter of solution immediately before use
STOCK: 1 M Tris-HCl, pH 7.7	
Buffer A: 0.1 M Tris-HCl, pH 7.7, 100 µM PMSF	
Buffer B: 0.1 M Tris-HCl, pH 7.7, 0.5 M KCl, 100 µM PMSF	
Dialysis tubing with a 10,000 Da molecular weight cut off (MWCO)	E.g., Spectra/Por® Membrane MWCO: 10,000, Spectrum Laboratories, Inc.
$(NH_4)_2SO_4$ (ammonium sulfate)	
Sorvall centrifuge	Or equivalent centrifuge/rotor combination to achieve the required gravitational force

Table 7.2 Chemicals, enzymes, and equipment for concentration of enzyme extracts

Material description	Comment
Amicon ultra centrifugal filters (0.5 ml, 10 kDa MWCO)	Millipore Amicon Ref. No. UFC501024 24Pk
Refrigerated (4°C) microcentrifuge	E.g., Eppendorf 5415R

Table 7.3 Chemicals, enzymes, and equipment for the test of mismatch cleavage activity

Material description	Comment
Thermocycler	E.g., Biorad C1000 Thermal cycler
PCR tubes	Life Science No 781340
TaKaRa Ex Taq™ polymerase (5 U/µl)	TaKaRa
10 Ex Taq™ reaction buffer	TaKaRa
dNTP mixture (2.5 mM of each dNTP)	TaKaRa
Agarose gel equipment	Horizontal electrophoresis from any general laboratory supplier

2. Add 1 M Tris-HCl (pH 7.7) and 100 mM PMSF to a final concentration of buffer A (0.1 M Tris-HCl and 100 µM PMSF) (NOTE: stocks and water should be kept at 4 °C, perform subsequent steps at 4 °C).
3. Centrifuge for 20 min at 2,600 × *g* in Sorvall GSA rotor or equivalent to pellet debris and transfer the supernatant to a clean beaker.
4. Bring the supernatant to 25 % ammonium sulfate (add 144 g/l of solution). Mix gently at 4 °C (cold room) for 30 min.

5. Centrifuge for 40 min at 4 °C at ~14,000 × g in Sorvall GSA rotor (~9,000 rpm) or equivalent. Discard the pellet.
6. Bring the supernatant to 80 % ammonium sulfate (add 390 g/l of solution). Mix gently at 4 °C for 30 min using a magnetic stir bar and plate.
7. Centrifuge for 1.5 h at 4 °C at ~14,000 × g. SAVE the pellet. Discard the supernatant (NOTE: Be careful not to disturb pellet while decanting the supernatant).
8. OPTIONAL: Pellets can be frozen at −80 °C for months.
9. Resuspend the pellets by vortexing in ~1/10 the starting volume with Buffer B (frozen pellets of the weed juice extract were suspended in 15 ml Buffer B and pellets of the celery juice extract in 10 ml Buffer B). Ensure that the pellet is thoroughly resuspended by pipetting it up and down or by vortexing.
10. Place the suspension into treated dialysis tubing use e.g. Spectra/Por® 7 10 kDa MWCO tubing (NOTE: follow manufacturer's guidelines for treatment of tubing before use).
11. Dialyze for 1 h against Buffer B at 4 °C with constant agitation of the buffer using a magnetic stir bar and plate. Use at least 2 l of buffer per 10 ml of suspended solution.
12. Repeat for a total of four steps with a minimum of 4 h dialysis for each step (NOTE: Longer dialysis is better, and it is often convenient to perform the third treatment overnight).
13. Remove the liquid from dialysis tubing. It is convenient to store ~75 % of the liquid in a single tube at −20 or −80 °C and the remainder in small aliquots for testing. This protein mixture does not require storage in glycerol and remains stable through multiple freeze–thaw cycles; however, limiting freeze–thaw cycles to five reduces the chance of diminished enzyme activity.
14. Perform activity test (Step 7.2.3, or proceed immediately to enzyme concentration, Step 7.2.2).

7.2.2 Concentration of Enzymes Using Amicon Ultra 10 kDa MWCO Centrifugal Filter Devices (for 0.5 ml Starting Volume; in 1.5-ml Tubes)

1. Perform with 600 µl of protein extract after dialysis.
2. Clear extract of plant material by centrifugation for 30 min at 10,000 × g, 4 °C.
3. Transfer 500 µl of the (cleared) supernatant to a filter device and keep the rest of the supernatant as the "before concentration" control.
4. Centrifuge the filter device with a collection tube inserted, as per the manufacturer's instructions for 30 min at 14,000 × g, 4 °C.
5. Remove the filter device, invert, and place in a new collection tube.
6. Centrifuge for 2 min at 1,000 × g, 4 °C.
7. Measure the recovered volume. This is your concentrated protein. Calculate the concentration factor with the following formula: concentration factor = starting volume/final volume.

7.2.3   Test of Mismatch Cleavage Activity

1. Produce TILLING-PCR products for mismatch cleavage tests with the concentrated enzyme extracts. The example below is for barley.
   GENES/PRIMER: nb2-rdg2a (1,500-bp PCR product)

| nb2-rdg2a_F2 | TCCACTACCCGAAAGGCACTCAGCTAC |
| nb2-rdg2a_R2 | GCAATGCAATGCTCTTACTGACGCAAA |

TILLING PCR REACTIONS (TaKaRa ExTaq enzyme):
Total volume: 25 µl

$10\times$ Ex Taq buffer (TaKaRa)	2.5 µl
dNTP mix (2.5 mM)	2.0 µl
Primer forward (10 µM)	0.3 µl
Primer reverse (10 µM)	0.3 µl
TaKaRa Taq (5 U/µl)	0.1 µl
Barley genomic DNA (5 ng/µl)	5.0 µl
H_2O (to 25 µl)	14.8 µl

2. TILLING PCR cycling program:
   95 °C for 2 min; loop 1 for 8 cycles (94 °C for 20 s, 73 °C for 30 s, reduce temperature 1 °C per cycle, ramp to 72 °C at 0.5 °C/s, 72 °C for 1 min); loop 2 for 45 cycles (94 °C for 20 s, 65 °C for 30 s, ramp to 72 °C at 0.5 °C/s, 72 °C for 1 min); 72 °C for 5 min; 99 °C for 10 min; loop 3 for 70 cycles (70 °C for 20 s, reduce temperature 0.3 °C per cycle); hold at 8 °C.
3. Mix 10 µl of PCR product with 10 µl weed digestion mix to a volume of 20 µl.
4. Incubate at 45 °C for 15 min.
5. Add 2.5 µl of 0.5 M EDTA (pH 8.0)—to stop the reaction.
6. Load a 10 µl aliquot on an agarose gel.
7. Analyze the samples by electrophoresis using a 1.5 % agarose gel. See Chap. 8 for example data.

Reference

Till BJ, Burtner C, Comai L, Henikoff S (2004) Mismatch cleavage by single-strand specific nucleases. Nucleic Acids Res 32:2632–2641

Chapter 8
Example Data

Abstract Standard agarose gel electrophoresis is a quick method for the evaluation of the quality and quantity of DNA. This chapter provides examples of genomic DNA produced using the low-cost extraction protocol, PCR amplification using the extracted genomic DNA, and enzymatic mismatch cleavage of PCR products with crude celery juice extract and weed juice extract to detect mutations.

8.1 Quality of Genomic DNA Obtained by Silica Powder-Based DNA Extraction Method

While spectrophotometric approaches (e.g., Nanodrop) provide a quick and accurate measure of DNA concentration and protein contamination, and fluorometric methods (e.g., Qubit) provide high sensitivity, it is advisable that when optimizing the DNA extraction protocol, samples are also run on a traditional agarose gel. This allows an estimation of DNA concentration (relative to concentration standards such as lambda DNA, Till et al. 2007), the extent of RNA carryover, as well as an estimation of the extent of DNA degradation, something which cannot be easily determined from the other two techniques (Figs. 8.1 and 8.2). Furthermore, chaotropic salts can lower the accuracy of spectrophotometric methods. Reducing sample degradation may be a key optimization step for some species. Alternative buffers can be employed to limit degradation and the copurification of secondary metabolites that can inhibit downstream molecular assays (see Sect. 8.2).

B.J. Till et al., *Low-Cost Methods for Molecular Characterization of Mutant Plants*, DOI 10.1007/978-3-319-16259-1_8

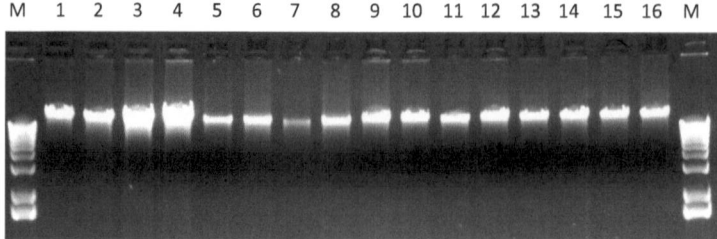

Fig. 8.1 Quality of barley genomic DNA extractions using combinations of self-made and commercial products. Eight microliters of each genomic DNA extraction was electrophoresed on a 0.7 % agarose gel. M = 1 kb Plus DNA ladder (Life Technologies). Lanes 1–4 are samples prepared with the DNeasy kit from Qiagen. Lanes 5–8 are samples prepared using DNeasy columns but with self-made 6M Guanidine thiocyanate buffer replacing commercial buffer AP3/E. Lanes 9–12 are samples prepared with self-made lysis buffer but with commercial DNA binding buffer AP3/E. Lanes 13–16 represent samples prepared using only self-made buffers described in Chap. 4

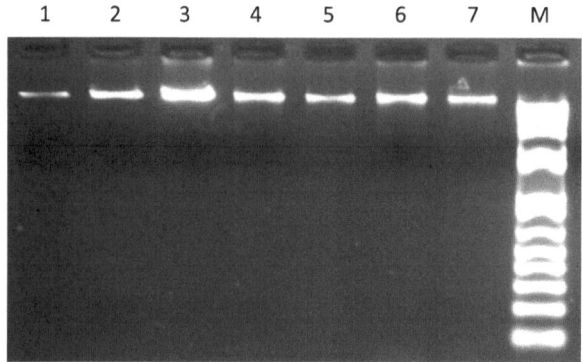

Fig. 8.2 Genomic DNA samples produced at the FAO/IAEA 2013 training course on "Plant Mutation Breeding: Mutation Induction, Mutation Detection, and Pre-Breeding." Lanes 1–3 represent lambda DNA concentration standards of 3, 10, and 30 ng/μL, respectively. Lanes 4–7 represent genomic DNA prepared by Ms. Sasanti Wisiarsih and Mr. Wijaya Murti Indriatama of Indonesia using the protocol described in Chap. 4 using a 6M KI buffer

8.2   Quality of Genomic DNA Obtained by Silica Powder-Based DNA Extraction Method Using Alternative Buffers

Alternative buffers can be used to extract DNA from tissues or species where the use of the standard buffer produces low-quality DNA. Tissue from *Sorghum bicolor* and grapevine *(Vitis vinifera)* were used to compare different buffer compositions (Tables 8.1 and 8.2).

Quality and quantity of genomic DNA using different buffers were assayed by native agarose gel electrophoresis (Fig. 8.3).

Table 8.1 Sorghum (*Sorghum bicolor*, Sb) genomic DNA extractions using the silica powder method with four different lysis buffers

Sample designation	Sb1a	Sb1b	Sb2a	Sb2b	Sb3a	Sb3b	Sb4a	Sb4b
Lysis buffer (LB)[a]	LB 1		LB 2		LB 3		LB 4	
Incubation temperature (for lysis)	RT	65 °C	RT	65 °C	RT	65 °C	RT	65 °C
DNA concentration (ng/µl)	11	15	2	3	6	5	7	7
Total yield (µg)	2.0	2.6	0.4	0.6	1.0	0.9	1.3	1.3

[a]See Table 4.3 for composition of buffers

Table 8.2 Grapevine (*Vitis vinifera*, Vv) genomic DNA extractions using the silica powder method with four different lysis buffers

Sample designation	Vv1a	Vv1b	Vv2a	Vv2b	Vv3a	Vv3b	Vv4a	Vv4b
Lysis buffer (LB)[a]	LB 1		LB 2		LB 3		LB 4	
Incubation temperature (for lysis)	RT	65 °C	RT	65 °C	RT	65 °C	RT	65 °C
DNA concentration (ng/µl)	22	32	7	3	11	6	4	9
Total yield (µg)	3.9	5.7	1.3	0.6	2.1	1.1	0.6	1.5

[a]See Table 4.3 for composition of buffers

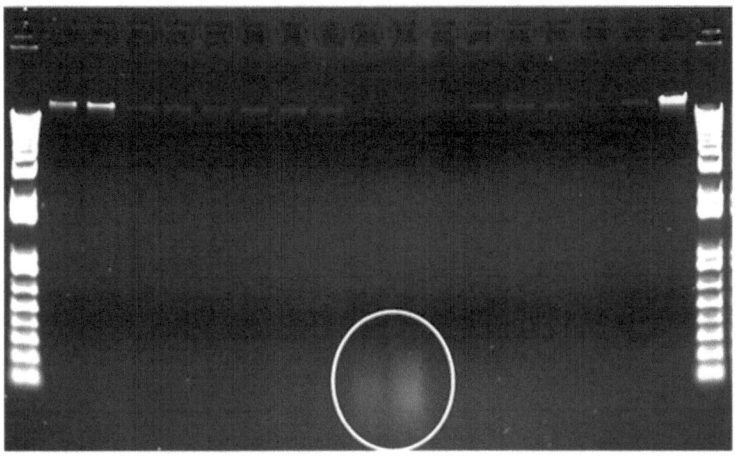

Fig. 8.3 Quality of genomic DNA extracted from sorghum and grapevine using the silica powder method with four different lysis buffers. Lanes 1–8 represent samples extracted from sorghum and lanes 9–16 from grapevine. Lanes 1, 2, 9, and 10 were prepared with lysis buffer 1 from Table 4.3. Lanes 3, 4, 11, and 12 with lysis buffer 2, lanes 5, 6, 13, and 14 with lysis buffer 3 and lanes 7, 8, 15, and 16 with lysis buffer 4. Lysis buffer 1 produced the highest yield with sorghum, but only degraded DNA with grapevine tissue (*circled*)

8.2.1   Summary

High yield and high quality genomic DNA can be recovered from sorghum using a simple lysis buffer. This buffer, however, is not suitable for grapevine DNA extraction and alternative buffers are required to recover high molecular weight DNA, albeit at a lower concentration than can be achieved from sorghum samples. This suggests further parameter changes can be made to increase yields.

8.3   Example of PCR Products Using TILLING Primers with Source Genomic DNA from a Commercial Kit and Low-Cost Silica Method

High quality and quantity of gene-specific PCR products are produced in reactions where source genomic DNA is extracted using either commercial kits or the low-cost silica method (Fig. 8.4).

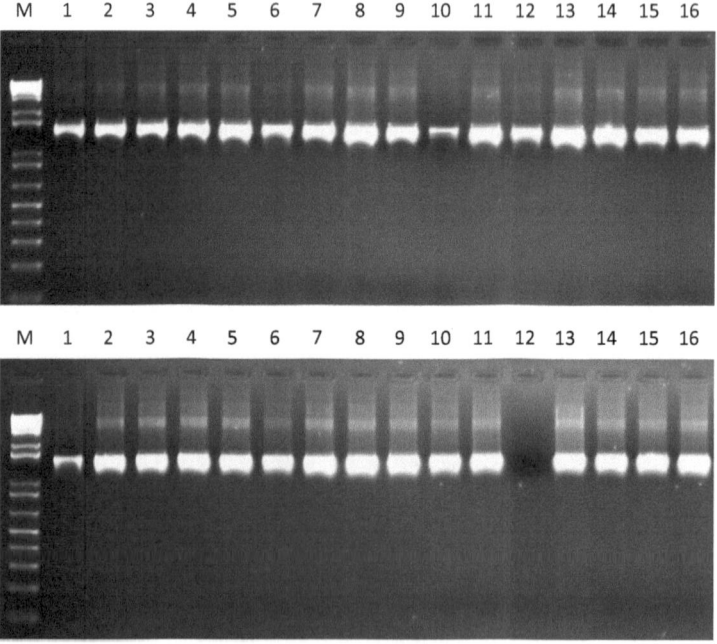

Fig. 8.4 PCR amplification of genomic DNAs described in Chap. 7 using primers for the barley nb2-rdg2a (*top panel*) and nbs3-rdg2a (*bottom panel*) gene targets as described in Hofinger et al. (2013). Samples are loaded in the same order as in Fig. 8.1

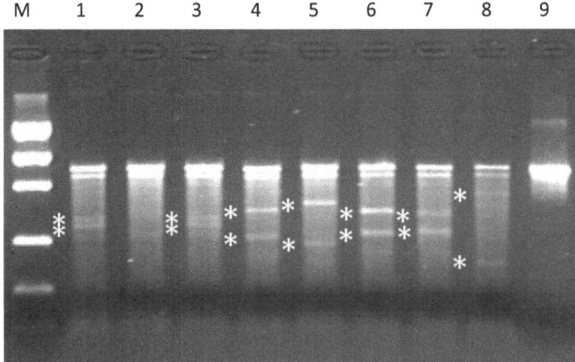

Fig. 8.5 Gel image of mutation discovery using crude celery juice extract for enzymatic mismatch cleavage. DNA from Arabidopsis plants with previously characterized induced point mutations in the *OXI1* gene was used (Till et al. 2004). Lane 2 represents a wild-type sample with no mutation. Lane 9 represents undigested PCR product. All other lanes contain samples with known mutations in the amplified region. Lower molecular weight bands representing cleavage products at the site of mutation are observable in all other lanes (marked by *asterisks*). This image was produced at the 2009 FAO/IAEA International Training Course on Novel Biotechnologies for Enhancing Mutation Induction Efficiency by Mr. Saad Alzahrani of Saudi Arabia, and Mr Azhar Bin Mohamad of Malaysia

8.4   Example of Low-Cost Agarose Gel-Based TILLING Assays for the Discovery of Induced Point Mutations

PCR products are the substrate for enzymatic mismatch cleavage assays for mutation discovery. Agarose gels provide a low-cost platform for mutation discovery using self-extracted enzymes (Fig. 8.5).

8.5   Example of Enzyme Activity Recovered from Weeds Compared to Crude Celery Juice Extract

The recovery of proteins from collected weeds versus celery is listed in Table 8.3. This was used to prepare reaction mixes (Table 8.4) to test for enzymatic activity. Samples were evaluated via standard agarose gel electrophoresis (Fig. 8.6).

8.5.1   Summary

Crude enzyme extracts of weeds show a similar activity to that of celery extract for the cleavage of single nucleotide polymorphisms. The per unit activity, however, was lower than for CEL I, likely owing to the co-precipitation of other plant

Table 8.3 Concentrations of protein extracts

Enzyme extract	Recovered volume	Concentration factor
Weed	~42 µl	11.9×
Celery	~33 µl	15.2×

Calculations of concentration factors after centrifugation with Amicon Ultra 10 kDa—starting volume: 500 µl ("Before" centrifugation is considered as 1× concentrated)

Table 8.4 Mismatch digestions using celery and weed enzyme extracts (prior and post centrifugation with Amicon Ultra 10 kDa filter devices

	Prior	Post-1	Post-2	Post-3
Enzyme	3.5 µl	0.5 µl	3 µl	6 µl
CEL I buffer	1.5 µl	1.5 µl	1.5 µl	1.5 µl
H$_2$O	5 µl	8.0 µl	5.5 µl	2.5 µl
Total volume	10 µl	10 µl	10 µl	10 µl

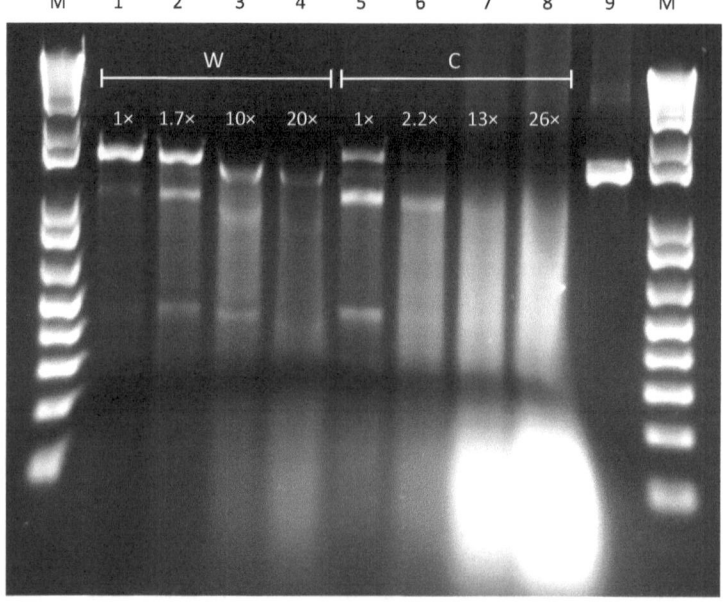

Fig. 8.6 Mismatch cleavage with crude enzyme extracts containing single-strand-specific nucleases prepared from weedy plants (W) or celery (C). PCR products of the target gene nb2-rdg2a (1,500-bp-PCR product) were produced from genomic DNA of barley containing a known SNP (Hofinger et al. 2013). The PCR products were digested with weed and celery enzyme extracts at different concentrations (listed above sample). Lower molecular weight bands are cleavage products. Cleavage activity from weed extract at 1.7× is similar to 1× activity observed from celery extracts. Undigested PCR product is loaded in lane 9

proteins in weeds, presumably including a larger amount of RuBisCO. This limitation can be overcome through the use of a simple centrifugation-based protein concentration step. Using this protocol, 150 ml of weed extract produces sufficient enzyme for approximately 2,000 reactions.

References

Hofinger BJ, Huynh OA, Jankowicz-Cieslak J, Müller A, Otto I, Kumlehn J, Till BJ (2013) Validation of doubled haploid plants by enzymatic mismatch cleavage. Plant Methods 9(1):43

Till BJ, Burtner C, Comai L, Henikoff S (2004) Mismatch cleavage by single-strand specific nucleases. Nucleic Acids Res 32:2632–2641

Till BJ, Cooper CJ, Tai TH, Colowit P, Greene EA, Henikoff S, Comai L (2007) Discovery of chemically induced mutations in rice by TILLING. BMC Plant Biol 7:19

Chapter 9
Conclusions

The approaches described here provide rapid and low-cost alternatives for sample preparation, genomic DNA extraction, and mutation discovery. When evaluating the methods, it is important to remember that protocol adaptations may be necessary to compensate for sample differences (species and genotype), environmental conditions in the laboratory, and quality of the water and chemicals used. Cost savings in DNA preparation must be balanced with the shelf-life and suitability of the samples for use in downstream applications. With the appropriate validation of sample quality and longevity, the protocols described here can provide sufficient DNA for a variety of molecular applications such as marker studies and TILLING, at approximately one tenth of the cost per sample when compared to commercial kits.

© International Atomic Energy Agency 2015   35
B.J. Till et al., *Low-Cost Methods for Molecular Characterization of Mutant Plants*,
DOI 10.1007/978-3-319-16259-1_9